D1284833

DOES THE ATOM HAVE A DESIGNER?

Lakhi N. Goenka, Ph.D.

December 2013
Updated July 2019

Lakhi N. Goenka, Ph.D.

1st Edition

ISBN-10: 0692789170
ISBN-13: 978-0692789179

Does the Atom Have a Designer?

We are in the position of a little child, entering a huge library whose walls are covered to the ceiling with books in many different languages. The child knows that someone must have written those books. It does not know who or how. It does not understand the languages in which they are written. The child notes a definite plan in the arrangement of the books, a mysterious order, which it does not comprehend, but only dimly suspects. That, it seems to me, is the attitude of the human mind, even the greatest and most cultured, toward God. We see a universe marvelously arranged, obeying certain laws, but we understand the laws only dimly. Our limited minds cannot grasp the mysterious force that sways the constellations.

- Dr. Albert Einstein

Table of Contents

Synopsis

Did the order and function of the Atom arise from random events?

"Does the Atom have a Designer?" critically examines if the order and complexity of the Atom arises simply from random events following the Big Bang. The Atom, with its complex structure, its quantized properties, and its many dynamic interactions, realizes the physical, electromagnetic, nuclear, chemical, and biological functionalities of the universe.

This book, written by a secular researcher searching for answers, examines subatomic behaviors such as wave-particle duality & quantum superposition, quark-gluon interactions within the atomic nucleus, and electron-photon interactions within atomic orbitals, as well as their role in realizing the many functionalities of the Atom.

The Why Questions related to the Atom are discussed in depth using Aristotle's four causes. The question: "Does your kitchen table have a Designer?" does not require a scientific or a mathematical explanation. It requires a logical one. Aristotle's theory of causality was developed to show that four related causes (or explanations) are needed to explain change in the world: a material, a formal, an efficient, and a final cause. A complete explanation of any material change will use all four causes. The Why Questions related to the Atom are investigated in great depth using this approach toward causality.

The commonly cited objection “Then who designed the Designer?” is also addressed in the book. The controversial and unverified Multiverse Hypothesis, often used against a Design argument, is also discussed.

And yes, your kitchen table *does* have a Designer.

(Note that this is an argument based on Design, and not on fine tuning.)

Atomic Structure

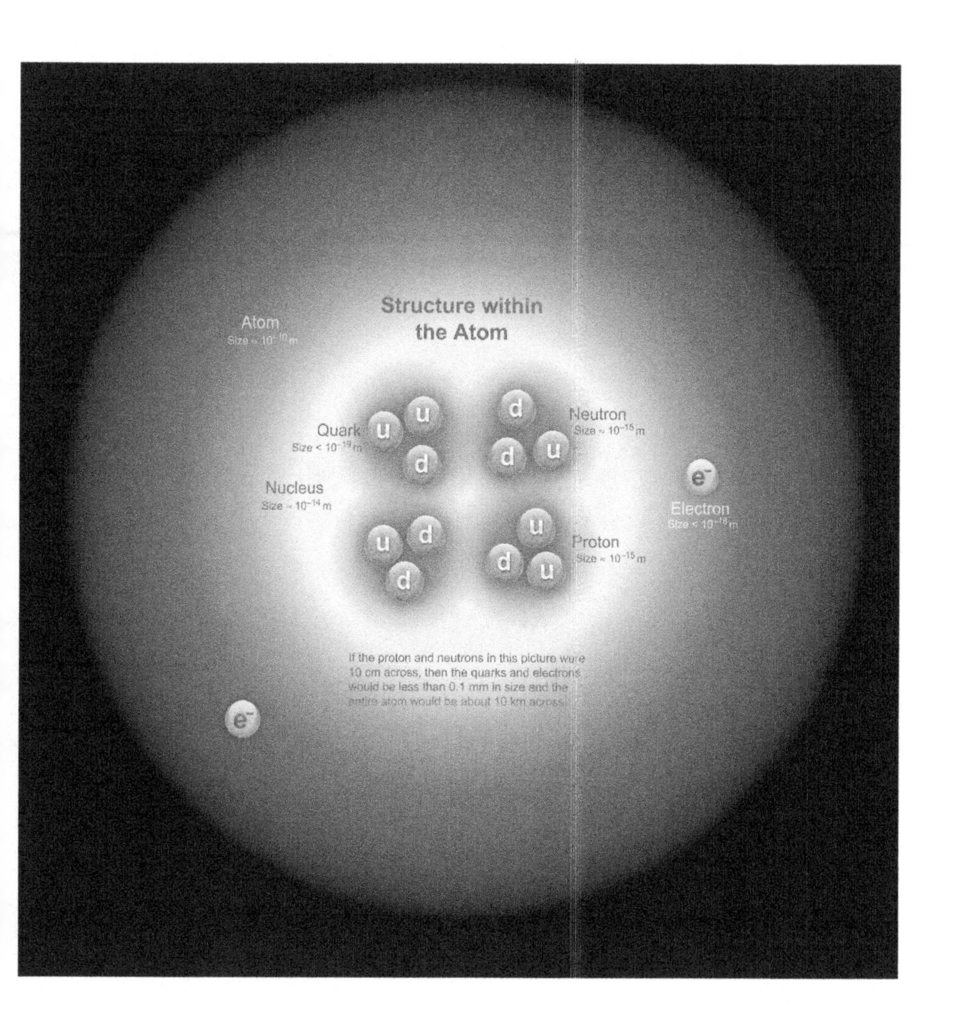

First-ever Image of a Hydrogen Atom

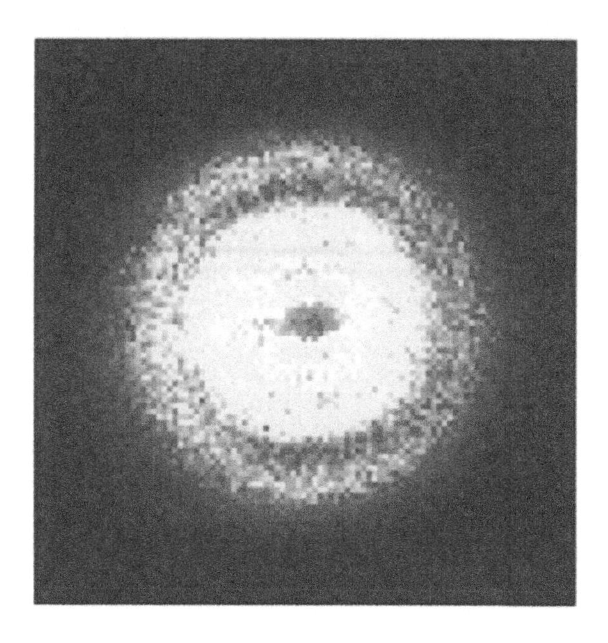

Using ultrafast lasers, wave packet experiments (2013) have been performed by A. S. Stodolna et al* illustrating how coherent superpositions of quantum mechanical stationary states describe electrons that move on periodic orbits around nuclei.
(Image courtesy *Physical Review Letters*)

* Related reference:
Hydrogen Atoms under Magnification: Direct Observation of the Nodal Structure of Stark States
A. S. Stodolna et al, Phys. Rev. Lett. **110**, 213001 – Published 20 May 2013

Introduction

The fundamental building block for the universe is not a passive billiard ball. It is a complex system of interacting particles called the Atom (which can perhaps be more appropriately referred to as The Atomic System). This book examines if the Atom with its complex structure and its multiple functionalities arises simply from random events, or if it has a Designer.

Even the simplest of atoms, Hydrogen and Helium, consist of numerous subatomic particles such as quarks, gluons, and leptons that interact together in complex ways. These subatomic particles represent a fine balance of forces, have special quantum properties, interact together in complex ways, follow complex laws, and obey multiple rules of order, all to ultimately provide function. Atoms don't simply follow laws—they provide function. Atoms are a fundamental system of parts (subatomic particles) that dynamically interact together to provide multiple levels of functionality.

The nucleus, or heart, of the Atomic System formed within fractions of a second after the Big Bang, and combined later with electrons that balanced out the charge within the system. The Atomic System represents the "starting level" of complexity for the universe, and consists of numerous subatomic particles whose quantum structure and interactions enable the physical, electromagnetic, nuclear, chemical, and biological functioning of the universe.

This book critically examines whether the Proton-Neutron-Electron structure of the Atom and its resulting functionalities are simply the result of random events. Topics investigated include an examination of the quantum properties and behaviors necessary for electromagnetism, the structure of and the interactions within the atomic nucleus, and the many resulting functionalities of the Atom. Related topics such as quantum field theories and the role of symmetry are also assessed.

Note that this book initially uses a particle-only formulation to discuss Atomic structure and its quantum interactions. This makes it easier to understand fundamental subatomic particles and their interactions, as well as the resulting atomic structure. However, Quantum Field Theory (QFT), in which particles are presumed to be excitations of underlying fields that permeate all of space, is gaining wider acceptance. QFT, which comes with some of its own issues, is discussed in depth later in the book.

The Why Questions about the Atom are addressed in some depth later in this book using Aristotle's four causes. Consider the question: "Does your kitchen table have a Designer?" The answer does not require a scientific or a mathematical explanation. It requires a logical one.

Another example of design would be if a hiker on a remote mountain trail comes upon some large rocks arranged into a message: "Have a great hike. There is an amazing vista a mile up the trail." The hiker would logically conclude that these rocks could not have arranged themselves into a message by chance, although the specific identity of the designer may never be established.

Aristotle's theory of causality was developed to show that there are four (related) causes (or explanations) needed to explain change in the world. A complete explanation of any material change will use all four causes.

Discussion of causes and causality go back to Ancient Philosophy, featuring prominently in Aristotle's *Metaphysics* and *Physics*. Ref [1] on Metaphysics discusses Aristotle's four causes in some depth.

From [1]: "The study of nature was a search for answers to the question "Why?" before and independently of Aristotle. A critical examination of the use of the language of causality by his predecessors, together with a careful study of natural phenomena, led Aristotle to elaborate a *theory* of causality. This theory is presented in its most general form in *Physics* II 3 and in *Metaphysics* V 5. In both texts, Aristotle argues that a final, formal, efficient or material cause can be given in answer to a why-question.

But Aristotle understood 'cause' in a much broader sense than we do today. In Aristotle's sense, a 'cause' or '*aiton*' is an explanatory condition of an object—an answer to a "why" question about the object. Aristotle classifies four such explanatory conditions—an object's form, matter, efficient cause, and teleology (the explanation of phenomena by the purpose they serve). In *Metaphysics*, an object's efficient cause is the cause that explains change or motion in an object. With the rise of modern physics in the seventeenth century, interest in efficient causal relations became acute, and it remains so today. And when contemporary philosophers discuss problems of causation, they typically mean this sense.

What is important is that this science consists in a causal investigation, that is, a search for the relevant causes. This helps us to understand why the most general presentation of Aristotle's theory of causality is repeated, in almost the same words, in *Physics* II 3 and in *Metaphysics* V 5. Although the *Physics* and the *Metaphysics* belong to two different theoretical enterprises, in both cases we are expected to embark on an investigation that will eventuate in causal knowledge, and this is not possible without a firm grasp of the interrelations between the four (types of) causes."

The commonly cited objection "Then who designed the Designer?" is also addressed later in the book.

The controversial and unverified Multiverse Hypothesis, often used against a Design argument, is also discussed in some depth. Princeton cosmologist Dr. Paul Steinhardt theorizes that multiverse theories have gained traction mostly because too much has been invested in theories that have failed (e.g., Inflation Theory and String Theory). He sees in them an attempt to redefine the values of science, to which he objects even more strongly: "A Theory of Anything (which he sees resulting from a Multiverse Hypothesis) is useless because it does not rule out any possibility and worthless because it submits to no do-or-die tests."

"Does the Atom have a Designer?" is a groundbreaking, first-of-its-kind analysis that critically examines if the Atom, with its Proton-Neutron-Electron structure and its multiple functionalities, simply arises from random events following the Big Bang.

Note that this is an argument based on Design, and not on Fine Tuning. Fine Tuning pertains to tuning an existing object or system by adjusting its parameters (or natural constants). Design pertains to the factors involved in the creation/design of an object or system, including its functionality.

And, by the way, your kitchen table *does* have a Designer.

Atomic Rules of Order

Particles within the Atom don't simply follow the conservation laws of mass, momentum, and energy; they also obey complex Rules of Order. These Rules of Order include Quantum Superposition, in which particles such as electrons, protons and neutrons can exist in multiple states of motion at the same time. They obey the Pauli Exclusion Principle, which states that no two identical particles with non-integer spin can exist in the same state of motion at the same time, and without which the Periodic Table of the elements could not exist. They follow Conservation Rules (in addition to conservation of energy and charge), in which quantum quantities such as the Baryon number and Lepton number are conserved. They exhibit Wave-Particle Duality, where particles can exhibit both particle-like and wave-like behavior. They exhibit probabilistic behavior quantified by the Schroedinger wavefunction. Subatomic particles such as electrons have intrinsic properties such as Quantum Spin, which is still not fully understood.

Note that field theories are mathematical entities being increasingly used to explain the wave-particle duality of subatomic particles. Particles are viewed as disturbances in fields that permeate all of space.

Matter and Energy

What are these subatomic particles *made of*? Amazingly, according to Physics, these particles are made from Energy. The mass of these particles results from asymmetric interactions with the recently discovered Higgs Boson, whose energy field is postulated to give other particles their mass. But we have *no fundamental understanding* of what Energy itself is.

Further, we have no fundamental understanding of what produces the wave-like and particle-like behaviors of Energy and its quantization into subatomic particles.

This can be explained using a quantum field formulation, but then we have no fundamental understanding of what created these quantized fields.

Einstein's equation $E= m*c^2$ quantifies the relationship between Mass and Energy, but the equivalence between Mass and Energy is subject to debate and discussion [2].

The Quantization of Matter (and Energy)

The Standard Model of Physics has been used successfully to predict many of the observed interactions of the quantum world. The Standard Model is a non-abelian gauge theory with the symmetry group $U(1)\times SU(2)\times SU(3)$ and has a total of twelve gauge bosons: the photon, three weak bosons and eight gluons.

Note that a gauge theory is a type of field theory in which the Lagrangian is invariant under a continuous group of local transformations. Further, the Special Unitary Group of degree n, denoted SU(n), is the Lie Group of $n\times n$ unitary matrices with determinant 1. The Special Unitary Group is a subgroup of the Unitary Group U(n), consisting of all $n\times n$ unitary matrices. A Lie Group is a smooth manifold obeying the group properties and that satisfies the additional condition that the group operations are differentiable.

Many of the limitations of the Standard Model of Physics are well known. Notable is the fact that it does not include Gravity, or Dark Matter. Further, the Standard Model has no fundamental understanding about what determines the mass of the different subatomic particles, or why their masses differ so widely. Some of the latest theories in Physics, such as String Theory and Supersymmetry, hope to address at least a few of these questions, but have not been experimentally verified.

But perhaps The Elephant in the Room is the question: Why did these quantized bundles of Energy we call Subatomic Particles form in the first place within moments after the Big Bang? Why are Matter and Energy, and their interactions, quantized to begin with? What *makes* Matter and Energy, and their interactions, quantized? Does Energy somehow interact with SpaceTime to create its quantization?

(Note that in a quantum field formulation, these quantizations occur because of the presence of underlying quantized fields that permeate through space. But then we have no fundamental understanding of how these quantized fields were created.)

Physics does not have an answer. Instead, Physics begins with experimental observations and attempts to fit them into a model called the Standard Model.

Perhaps a better question would be: What would be the effect if Matter and Energy, and their interactions, were *not* quantized? The answer jumps out at us: We would not have Atoms. Without the Atom, we would not have all the Elements that make up the Universe. We would not have the multiple functionalities necessary for the functioning of the Universe (see later section on Atomic Functionality and Purpose). The Universe could not function without the Atom. The Universe could not function without the Quantization of Matter and Energy, and of their interactions.

Charge, Spin, and Electromagnetism

Charge and Spin are two fundamental properties (in addition to Mass) that distinguish one subatomic particle from another. If subatomic particles are ultimately made from Energy, what then creates the fundamental quantum properties we call Charge and Spin? The occurrence of such properties can be somewhat explained by the underlying symmetry of the physical laws governing their interaction. This symmetry then requires that certain fundamental properties (that we call Charge and Spin) must be conserved. But this does not tell us what Charge and Spin are, or what enables them. In other words, symmetry underlying the physical laws is a necessary condition, but symmetry alone is not sufficient to fully explain the creation of conserved properties such as Charge and Spin. We know from experience how Charge behaves, but we have *no fundamental understanding* of what Charge is, or what *makes* it behave in this way. The same is true for Spin. Note that Charge and Spin are two fundamental properties that, along with Energy and perhaps SpaceTime, enable the creation of all Electricity, Magnetism, and Electromagnetism.

What causes the force between particles?

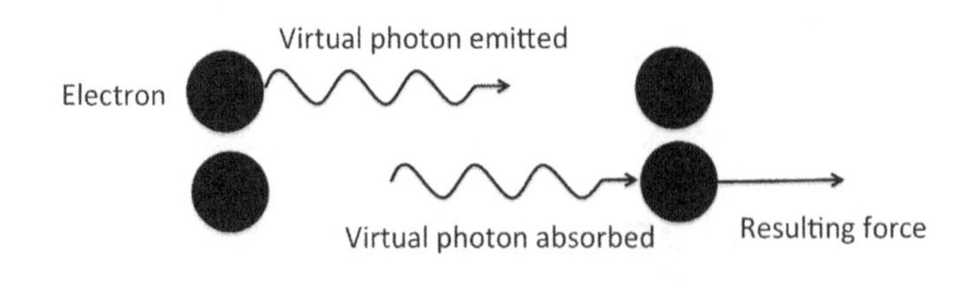

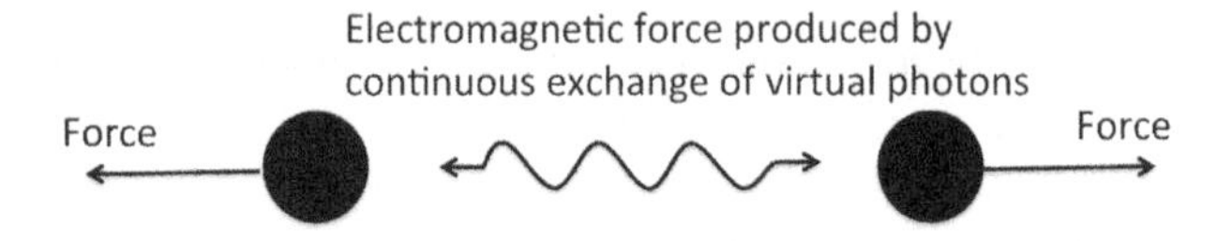

The interaction of two charged particles occurs through the creation of "virtual" photons that one particle emits toward the other. What *enables* the creation of such virtual photons? What *causes* their interactions between charged particles?

Further, when we say a particle has Charge, it really means that the charged particle *has the ability* to emit and to absorb virtual Photons which travel at the speed of light and with a certain frequency and amplitude, and to thereby interact with other particles with Charge. What *gives* it this ability? (Note that another way of stating this is that the Charged Particle has *the ability* to emit (create) and to interact with an Electromagnetic Field.) And could this simply be the result of random events?

The Electron, as it drops from a higher to a lower atomic orbital, emits a Photon. The Electron has *the ability* to convert Mechanical Energy to Electromagnetic Energy.

The mechanism by which a Photon, which in itself has no Charge, can transmit the electromagnetic force between charged particles remains unclear. Note that Light is an Electromagnetic wave, while quantized Light particles (Photons) have no electrical Charge. This adds to the underlying complexity of the Photon as well as perhaps charged particles such as Electrons and Quarks.

Photons have no rest mass. This is *necessary* for the functionality of the Atomic System so that it does not constantly lose or gain mass as orbiting Electrons within the Atom emit and absorb Photons. (Note that this is not the case for gluons within the Atomic Nucleus as they are always confined within it.)

Why is the Photon massless? Why does it not interact with the Higgs Field? What *makes* it massless? Having zero mass is *necessary* for Photons to travel at the Speed of Light.

Interestingly, from the reference frame of a traveling Photon, Space and Time shrink down to nothing.

Note that the corresponding field theory is called Quantum Electrodynamics (QED), which is the study of the interaction of the electromagnetic field with charged particles. Photons are excitations that travel through the underlying electromagnetic field and can be depicted as localized wave packets. Field theories seek to resolve the wave-particle duality exhibited by subatomic particles.

Wave-Particle Duality and Schroedinger's Wavefunction

The wave-particle and probabilistic behavior of subatomic particles can be described by Edwin Schroedinger's Wavefunction, which is *a semi-empirically derived* mathematical equation. It assumes a wave-particle nature of subatomic particles, energy conservation, as well as Einstein's famous equation E= m*c^2 and Planck's postulate E= h*f (the energy of an emitted Photon is proportional to its frequency). But it's worth noting that while the behavior of surface waves on water can be explained by the time-dependent interaction of pressure and fluid momentum, there is no such corresponding analogy to explain the wavelike behavior of light quanta (Photons) or the other subatomic particles. Schoredinger's Wavefunction is used to quantify the probabilities of occurrence of different quantum states.

Time-dependent Schrödinger equation
(single non-relativistic particle)

$$i\hbar\frac{\partial}{\partial t}\Psi(\mathbf{r},t) = \left[\frac{-\hbar^2}{2\mu}\nabla^2 + V(\mathbf{r},t)\right]\Psi(\mathbf{r},t)$$

Does the Atom Have a Designer?

The Pauli Exclusion Principle can be explained by Schroedinger's Wavefunction, which disallows the existence of two identical particle states for particles with half-integer spin within an Atom. No two electrons in an Atom can have identical quantum numbers. The Pauli Exclusion Principle is the result of one of our most basic observations of nature: Particles of half-integer spin have anti-symmetric wavefunctions, and particles of integer spin have symmetric wavefunctions.

Wavelike behavior is also exhibited by electrons moving around the nucleus, and without which they would have unstable orbits and crash into the atomic nucleus.

The wave-particle behavior of subatomic particles leads to the creation of the entire Periodic Table of the elements, as well as to the stability of the electron orbitals. What *makes* these particles exhibit this probabilistic wavelike behavior? What *enables* such behavior?

Further, the Wavefunction has both real and imaginary parts to it. Some electron orbitals, such as the "d" and "f" orbitals, are so complex that they project into 5-dimensional and 7-dimensional space. It is possible to map the "d" orbitals back into 3-dimensional space, but the result is a strange looking "dz^2" orbital [3].

The laws and rules governing the interactions of matter do not necessarily give us an understanding of *how* it does so, or what enables and causes such behavior. The laws that the subatomic particles follow can be used to predict the outcome of their interactions, but do not necessarily explain the details of the particle interactions that are necessary to produce such outcomes. They also do not explain the composition of the particles themselves that enables the particles to undergo such interactions and exhibit such behaviors. For example, when an electron drops to a lower orbital, it emits a particle called a photon that we see as light. But there is no explanation as to what *enables* the electron to emit another subatomic particle (the photon); nor is there any understanding of the *time-dependent separation* of the photon from the electron. How did this photon materialize from the electron? And how did it then travel as a wave at the speed of light at a certain frequency, direction, and amplitude? And how does this particle sometimes behave like a wave (akin to ripples in a pond)? How was this mechanism incorporated into the design of the electron and photon? *How was this fantastic machinery put in place?*

Further, the mechanism by which a Photon, which in itself has no Charge, can transmit the electromagnetic force between charged particles remains unclear. This adds to the underlying complexity of the Photon as well as perhaps of the other particles with Charge such as Electrons and Quarks.

In 1951, Albert Einstein wrote to his friend Michael Besso about the unfathomable nature of the photon:
All these fifty years of conscious brooding have left me no nearer to the answer to the question: What are light quanta? Nowadays every rascal thinks he knows the answer, but he is simply deluding himself.

How *can* something behave like both a particle and a wave? And are such complex behavioral mechanisms that are incorporated into the Photon, as well as the other subatomic particles, simply from random events?

As mentioned before, the latest thinking is that wave-particle behavior arises because of underlying quantized fields that permeate all of space. This is discussed in more depth later in the book in the Chapter entitled "Quantum Field Theory".

The Role of Symmetry

Some of the quantum interactions between different energy fields can be explained by constraints imposed by the underlying gauge symmetry. Gauge symmetry is used to unify Quantum Electrodynamics (QED) with Quantum Chromodynamics (QCD). (Note that QCD is a field theory that describes the interactions of particles effected by the strong nuclear force, discussed later in the book.)

A comprehensive discussion on the role of symmetries can be found in *Symmetry and Symmetry Breaking (2003)*, Stanford Encyclopedia of Philosophy [4]. To quote from this article: "In the language of modern science, the symmetry of geometrical figures -- such as the regular polygons and polyhedra -- is defined in terms of their invariance under specified groups of rotations and reflections....A different notion of symmetry emerged in the seventeenth century, grounded not on proportions but on an equality relation between elements that are opposed, such as the left and right parts of a figure. Crucially, the parts are *interchangeable with respect to the whole* -- they can be exchanged with one another while preserving the original figure. This latter notion of symmetry developed, via several steps, into the concept found today in modern science. One crucial stage was the introduction of specific mathematical operations, such as reflections, rotations, and translations, that are used to describe with precision how the parts are to be exchanged. As a result, we arrive at a definition of the symmetry of a geometrical figure in terms of its *invariance* when equal component parts are exchanged according to one of the specified operations....The next key step was the generalization of this notion to the group-theoretic definition of symmetry, which arose following the nineteenth-century development of the algebraic concept of a group, and the fact that the symmetry operations of a figure were found to satisfy the conditions for forming a group. ...The group-theoretic notion of symmetry is the one that has proven so successful in modern science. Note, however, that symmetry remains linked to beauty (regularity) and unity: by means of the symmetry transformations, distinct (but "equal" or, more generally, "equivalent") elements are related to each other and to the whole, thus forming a regular "unity". The way in which the regularity of the whole emerges is dictated by the nature of the specified transformation group. Summing up, a *unity of different and equal elements* is always associated with symmetry, in its ancient or modern sense; the way in which this unity is realized, on the one hand, and how the equal and different elements are chosen, on the other hand, determines the resulting symmetry and in what exactly it consists...

The definition of symmetry as "invariance under a specified group of transformations" allowed the concept to be applied much more widely, not only to spatial figures but also to abstract objects such as mathematical expressions — in particular, expressions of physical relevance such as dynamical equations.

...We may attribute specific symmetry properties to phenomena or to laws (*symmetry principles*). It is the application with respect to laws, rather than to objects or phenomena, that has become central to modern physics, as we will see. Second, we may derive specific consequences with regard to particular physical situations or phenomena on the basis of their symmetry properties (*symmetry arguments*).

...The rich variety of symmetries in modern physics means that questions concerning the status and significance of symmetries in physics are not easily addressed...Symmetries are seen as a substantial part of the physical world: the symmetries of theories represent properties existing in nature, or characterize the structure of the physical world. It might be claimed that the ontological status of symmetries provides the reason for the methodological success of symmetries in physics.

...While global continuous symmetries can be directly observed -- via such experiments as the Galilean ship experiment – a local symmetry can have only indirect empirical evidence. (In the Galilean ship experiment, Galileo urges us to consider the motion of objects inside the cabin of a moving ship. He claims that no experiments carried out inside the cabin, without reference to anything outside the ship, would enable us to tell whether the ship is at rest or moving smoothly across the surface of the Earth)...In theories with local gauge symmetry, the matter field is embedded in a gauge field, and the local symmetry is a property of both sets of fields jointly. Because of this, there is no analog of the Galilean ship experiment for local symmetry transformations.

...According to Wigner [5], the spatiotemporal invariance principles play the role of a prerequisite for very possibility of discovering the laws of nature: "if the correlations between events changed from day to day, and would be different for different points of space, it would be impossible to discover them". For Wigner, this conception of symmetry principles is essentially related to our ignorance (if we could directly know the laws of nature, we would not need to use symmetry principles in our search for them).

...To conclude, symmetries in physics have many interpretational possibilities, and how to understand the status and significance of physical symmetries clearly presents a challenge to both physicists and philosophers."

Note that the Standard Model of Physics cannot predict the values of 26 constants it contains (including the mass of all the subatomic particles) utilizing fundamental principles such as gauge symmetry, and instead must resort to experimentation to determine these. Note also that particle masses are not quantized, and do not observe any form of symmetry.

It's also interesting to note that without these underlying symmetries, and their breakings, many of the fundamental quantum particle behaviors and interactions within the Atom would not be possible. Without these underlying symmetries, and their breakings, the Atom (and the Universe) could not effectively function. These underlying symmetries, and their breakings, are *necessary* for the Atom, and for the Universe, to function.

Recent (2016) high-energy tests at the Large Hadron Collider in Europe have found no evidence of the Supersymmetric particles that are linked to String Theory (which assumes that the fundamental building blocks of matter are vibrating strands of energy). This puts into serious doubt the validity of String Theory with its assumptions of eleven-dimensional spaces and parallel universes, as well as its potential to explain Dark Matter, and leaves Physics at a loss to reach beyond the Standard Model.

The Atomic Nucleus

Further behavioral complexity can be found within the atomic nucleus, which consists of Protons and Neutrons (which are subsets of particles called Nucleons). Initial atomic nuclei were produced moments after the Big Bang. Each Proton and Neutron is made up three particles called Quarks that are held together, and interact with, eight particles called Gluons. (Note that the exact number of Gluons within the nucleus is murky, and has not yet been definitively determined. Also, note that Physicists today believe that at the super-high energy levels right after the Big Bang, the electromagnetic, the strong and the weak forces were one, but within moments separated out as the Universe rapidly expanded and energy levels decreased.)

The particles within the atomic nucleus are in constant motion, with Protons and Neutrons constantly “jiggling” as they interact with each other, while Quarks within the Protons and Neutrons are also constantly jiggling as they interact with the Gluons.

A property of Quarks, labeled Color, is an essential part of the make up of Quarks. The resulting three different Color Forces have extraordinary properties within the Proton and Neutron. The Color Force does not change as the separation distance between Quarks is increased (resulting in what is called *Quark Confinement*), and paradoxically, exerts little force at very short distances so that Quarks can interact like free particles when very close to each other. The term *Asymptotic Freedom* is often used to refer to the free interaction of Quarks with Gluons over small distances.

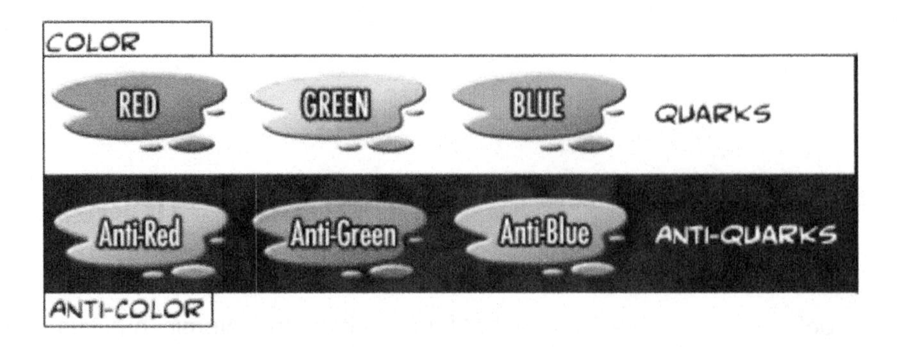

The behavior of these forces can be explained by the underlying chiral symmetry breaking of these interactions. It's interesting to note that strong-force interactions within the atomic nucleus would not be possible without the Asymptotic Freedom that was realized from this symmetry breaking.

The presence of quarks with like charges within the Proton and Neutron results in the production of strongly repulsive electrostatic forces. These repulsive forces are countered in part by the dynamic interactions of the strongly attractive nuclear forces within the Proton and Neutron. This dynamic interaction of push-pull forces is necessary to create nucleons with multiple quarks and gluons. It also enables a Nucleus with multiple Protons and Neutrons that can combine together to produce the different elements in the universe.

Three different Colors are necessary to allow two or more Quarks with otherwise identical quantum properties to exist together within a Nucleon. Since Quarks have a non-integer Spin, they would, in the absence of Color, violate the Pauli Exclusion Principle. But this then also implies that the Quarks within Protons and Neutrons display wavelike behavior. There is a whole field called Quantum Chromodynamics that describes these complex interactions. These strong interactions within the Atomic Nucleus result in stored potential nuclear energy that is released during nuclear fission and fusion reactions, and which fuel the stars (including our Sun).

Three Color forces are also necessary within the Atomic Nucleus to enable multiple adjacent particles to dynamically interact with each other to create the resulting three-dimensional structure of the Nucleus. This three-dimensional structure of the Atomic Nucleus which enables multiple Protons and Neutrons to combine and dynamically interact together would not be possible with only one Color force.

This unique structure of the Proton (with two up-Quarks, one down-Quark and eight Gluons) is necessary in the Proton-Proton cycle of a nuclear fusion reaction. At the beginning of this cycle, the up-Quarks within the Proton are converted to down-Quarks, ultimately resulting in the conversion of a Proton to a Neutron. This Neutron then combines with a Proton to form Deuterium nuclei. This cycle continues to ultimately create Helium-3 nuclei, while releasing vast amounts of energy. Without these fusion reactions, stars (and our Sun) could not exist. The nuclear fusion reaction in stars also results in the creation of heavier elements from lighter ones as the intense pressures and temperatures within them helps to bind protons and neutrons together. The Proton-Neutron structure of the Atomic Nucleus is necessary to realize multiple levels of functionality necessary for the functioning of the Universe. The Atomic Nucleus is a *purposeful* assembly of numerous dynamically-interacting subatomic particles.

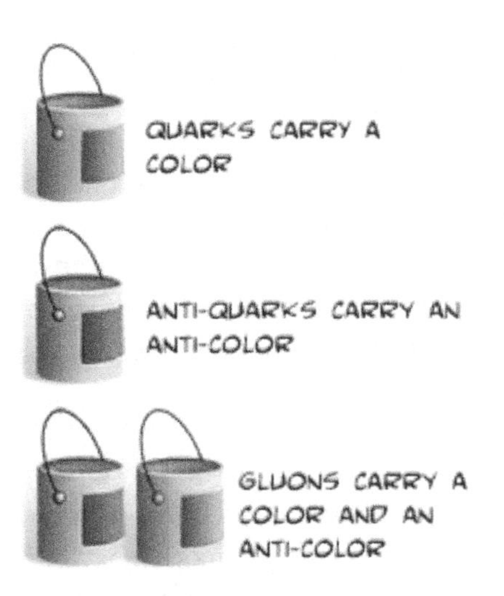

Neutrons consist of two down-Quarks and one up-Quark that interact with eight Gluons, and have no net charge. They help bind several Protons in the nucleus together as they, like Protons, contribute to the binding nuclear force within the nucleus, but do not repel each other like two positively-charged Protons would.

The complexity of particle interactions within Protons, as well as the limited understanding of such interactions, is illustrated by recent experiments investigating the contribution of Gluon Spin to the interactions. To quote from www.Physics.org (2014): “Results from experiments at the Relativistic Heavy Ion Collider (RHIC), a particle collider located at the U.S. Department of Energy's Brookhaven National Laboratory, reveal new insights about how quarks and gluons, the subatomic building blocks of protons, contribute to the proton's intrinsic angular momentum, a property more commonly known as "spin." Specifically, the findings show for the first time that gluons make a significant contribution to proton spin, and that transient "sea quarks"—which form primarily when gluons split—also play a role.

The new precision measurements will help solve a mystery that has puzzled physicists since the 1980s, when findings from early spin experiments in Europe and elsewhere simply didn't add up. Those experiments showed that the spins of quarks—including the three valence quarks that determine most of the basic properties of the proton—plus antiquarks could account for, at most, a third of the proton's total spin."

Note the reference to the formation of "transient sea quarks" during these interactions. This is another indication of the complexity of particle interactions within Protons, as well as within Neutrons and the Atomic Nucleus.

When combined within the nucleus, Neutrons dynamically interact with Protons through the creation and exchange of virtual Pions (consisting of a Quark-antiQuark pair) that materialize out of Nothing. During this interaction, Quarks within each Nucleon are dynamically changing color so that each Quark has a different color. This interaction produces the residual strong force (also referred to as the nuclear force) that binds Protons and Neutrons together within the atomic nucleus. Note that the electrostatic repulsive force between two positively-charged Protons varies inversely with the square of the separation distance between them, and increases exponentially as the Protons are close together within the Nucleus. The nuclear force produced by the Proton-Neutron interaction is strong and overcomes this repulsive electrostatic force.

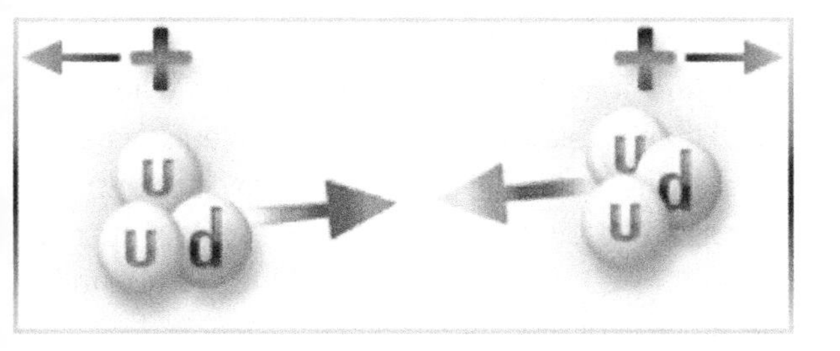

The complimentary structure of Protons and Neutrons is perfectly suited for them to combine and dynamically interact with each other within the Atomic Nucleus.

This interaction of virtual Pions with Protons and Neutrons within the nucleus is a bit like saying that a functional part of a rocket ship materializes from Nothing, interacts dynamically with existing functional parts of the rocket ship, and then disappears into Nothing.

(In the field formulation, the existence of virtual particles can be explained by the Uncertainty Principle, which states that it is impossible to know the position and momentum of a particle simultaneously with complete accuracy; or the energy involved in an interaction and the duration of that interaction. It is the latter that makes it possible for a virtual particle to come into existence from Nothing, borrowing energy and returning it within the time mandated by the uncertainty principle. But although this might be theoretically possible, we have no fundamental understanding of the *underlying physical mechanisms* necessary to make this happen.)

The complexity of such interactions is summed up by prominent physicist Richard Feynman, whose famous Feynman Diagrams are used to depict the interactions within the atomic nucleus: *If you think you understand quantum mechanics, you don't understand quantum mechanics.*

This Proton/Neutron structure of the atomic nucleus allows for the combination of vastly different numbers of Protons and Neutrons to produce all the elements in the universe (as diverse as Hydrogen, Nitrogen, Carbon, Iron, and Mercury). Heavier elements have more Neutrons than Protons than lighter ones to provide sufficient binding energy to keep the nucleus intact. And this structure of the nucleus also enables nuclear fission, during which the nucleus of a heavier atom is split to form smaller nuclei while releasing a huge amount of energy.

The three-dimensional structure of the nucleus also helps realize isotropic mechanical and thermal properties of the resulting element.

The Quantum Spin of particles within the nucleus results in a small magnetic field (referred to as the Nuclear Magnetic Moment). The Nuclear Magnetic Moment of each element is different. Further, the nuclear magnetic field changes depending on the proximity of other adjacent atoms. Atoms that are crowded together would produce magnetic fields that are different in strength from atoms that are spaced further apart. This property is used in Nuclear Magnetic Resonance to determine the types of atoms within the molecules being investigated.

This Proton/Neutron structure has a net positive charge that enables interactions with orbiting Electrons having an equal but opposite charge, and which then enables the creation of all of Electromagnetism. Electrons emit Photons as they drop from a higher-energy to a lower-energy orbital while dynamically interacting with the atomic nucleus.

Further, it is crucial for multiple Protons to be able to bond together within the Atomic Nucleus. Without multiple Protons, more than one Electron could not simultaneously orbit the nucleus. Without the resulting structure of the Atomic System with multiple orbiting Electrons, it would not be possible for several elements to chemically bond with each other to create increasingly complex chemicals and compounds, including carbohydrates, proteins, and the DNA molecule.

The multiple Protons (and Neutrons) in the nucleus, with the corresponding multiple orbiting Electrons, also determines the physical properties of the resulting element such as its density, thermal conductivity, and electrical conductivity, and categorizes it into an insulator, conductor, or semi-conductor. This structure of the Atom with multiple Protons and multiple orbiting Electrons also helps realize the entire electromagnetic spectrum.

The following criteria are necessary for the creation of the entire Electromagnetic Spectrum, as well as to realize the other functionalities of the Atom:

Electrons have (1) a negative charge and (2) half-integer spin. They (and the (3) massless Photons they emit) also need the ability to exhibit (4) wave-like and (5) particle-like behavior. Further, (6) they need to orbit and dynamically interact with the Atomic Nucleus (7) having a net positive charge and which has (8) the ability to combine different numbers of Protons and Neutrons.

Protons need to have (9) two up-Quarks and one down-Quark (10) interacting with eight Gluons (11) to produce three strong color forces and the resultant stored nuclear potential energy, while Neutrons need to have (12) one up-Quark and two down-Quarks (13) interacting with eight Gluons. This combination gives the Neutrons (14) a net zero charge. Further, (15) Protons and Neutrons dynamically interact with each other through (16) external virtual Pions consisting of a Quark – anti-Quark pair that (17) materializes from Nothing to (18) produce the residual nuclear force that enables multiple positively-charged Protons to combine together with Neutrons in the Atomic Nucleus.

Further, the negative charge of an Electron precisely balances out the positive charge of a Proton. And it is this fundamental structure of the Proton and the Neutron within atomic nuclei that enables nuclear fusion reactions that fuel the Stars (including our Sun), and which then also result in the creation of heavier elements from lighter ones.

Hydrogen, Deuterium, and Helium were the initial Atomic nuclei to form within moments after the Big Bang. The Hydrogen nucleus, which has one Proton, consists of 11 subatomic particles. Deuterium, with one Proton and one Neutron, would have 22 subatomic particles, while the Helium nucleus, which has two Protons and two Neutrons, consists of 44 subatomic particles. How were these particles able to *selectively combine* to form the nucleus of the Atom, and then dynamically interact to produce multiple levels of functionality?

Further, each Proton consists of two up-Quarks, both with a positive charge of +2/3, and a down-Quark with a negative charge of -1/3. How were these two positively-charged up-Quarks, which should repel each other, able to come together to combine into a Proton (with a resultant net positive charge of +1)? Similarly, how were two positively-charged Protons able to combine together into an Atomic Nucleus, without the necessary composition and properties of mysterious particles called Gluons to enable this?

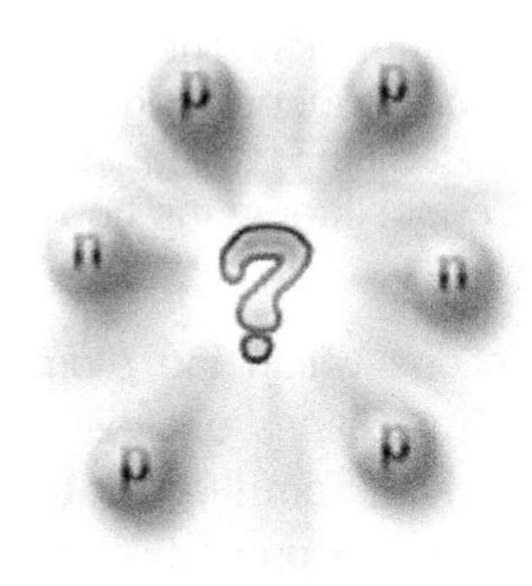

And what *makes* the virtual Pions that dynamically interact with existing Quarks and Gluons within the atomic nucleus appear out of Nothing? What is the underlying *cause* of this fantastic behavior, and the resulting Strong Force within the nucleus?

The complexity of such interactions and the immaturity of our understanding are graphically summarized by Albert Einstein:

We are in the position of a little child entering a huge library whose walls are covered to the ceiling with books in many different tongues. The child knows that someone must have written those books. It does not know who or how. It does not understand the languages in which they are written. The child notes a definite plan in the arrangement of the books, a mysterious order, which it does not comprehend, but only dimly suspects. That, it seems to me, is the attitude of the human mind, even the greatest and most cultured, toward God. We see a universe marvelously arranged, obeying certain laws, but we understand the laws only dimly. Our limited minds cannot grasp the mysterious force that sways the constellations.

As noted before, this book initially uses a particle-only formulation to discuss Atomic structure and its quantum interactions. This makes it easier to understand fundamental subatomic particles and their interactions, as well as the resulting atomic structure. However, Quantum Field Theory (QFT), in which particles are presumed to be excitations of underlying fields that permeate all of space, is gaining wider acceptance. QFT, which comes with some of its own issues, is discussed in depth later in the book.

The corresponding field theory to explain interactions within the atomic nucleus is called Quantum Chromodynamics (QCD). Gluons are excited quantizations of the underlying field. The name derives from the fact that (as discussed above) Quarks within the nucleus have a property called "Color" that determines how they interact.

Atomic Stability

Atoms are inherently stable systems. The Proton, which consists of three quarks that are held together by gluons, is one of the few composite particles that does not experience decay into other particles. Neutrons, however, do not experience decay only when they are combined with Protons within the atomic nucleus. However, when not within the nucleus, free Neutrons decay into a Proton, Electron, and an Antineutrino within about 15 minutes. The properties of both the Proton and Neutron ensure a stable atomic nucleus, without which the Atom could not exist as a stable system that stays intact for thousands of millions of years, and without which the universe could not function.

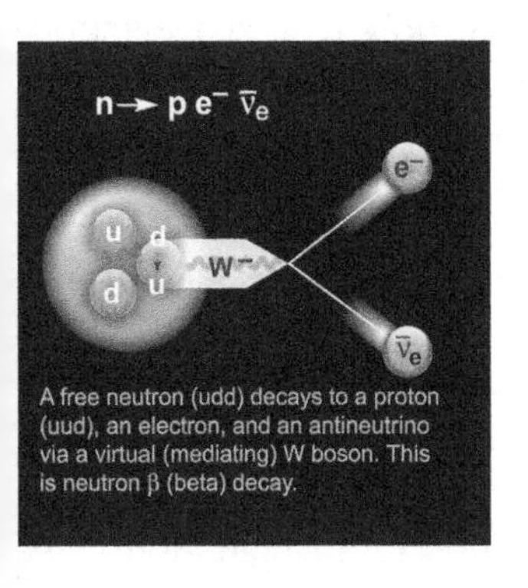

A free neutron (udd) decays to a proton (uud), an electron, and an antineutrino via a virtual (mediating) W boson. This is neutron β (beta) decay.

The wavelike behavior, as well as the quantization of charge, exhibited by electron orbitals around the atomic nucleus enables them to have stable orbits. According to classical physics, the constant acceleration experienced by an electron in a purely circular electron orbit would cause it continuously lose electromagnetic energy and thereby crash into the nucleus.

With the development of quantum mechanics and experimental findings (such as the two slits diffraction of electrons), it was found that the orbiting electrons around a nucleus could not be fully described as particles, but could only be explained by the wave-particle duality. The electrons do not orbit the nucleus in the sense of a planet orbiting the sun, but instead exist as standing waves. The lowest possible energy an electron can take is therefore analogous to the fundamental frequency of a wave on a string. Higher energy states are then similar to harmonics of the fundamental frequency.

The electrons are never in a single point location, although the probability of interacting with the electron at a single point can be found from the Schroedinger wave function of the electron.

Electrons also exhibit particle-like properties as they orbit the atomic nucleus. There is always an integer number of electrons orbiting the nucleus. Electrons jump between orbitals in a particle-like fashion. For example, if a single photon strikes the electrons, only a single electron changes states in response to the photon. They also retain particle like-properties; each wave state has the same electrical charge as the electron particle, and has a single discrete spin (spin up or spin down).

Thus, unlike planets revolving around the Sun, electrons cannot be described simply as solid particles. A more accurate analogy for electron orbitals might be that of a large and often oddly shaped "atmosphere" (the electron), distributed around the atomic nucleus. Atomic orbitals exactly describe the shape of this "atmosphere" only when a single electron is present in an atom. When more electrons are added to a single atom, the additional electrons tend to more evenly fill in a volume of space around the nucleus so that the resulting collection (sometimes termed the atom's "electron cloud") tends toward a zone of probability around the nucleus describing where the atom's electrons will be found.

The quantization of electromagnetism into photons, and the wave-particle duality as well as the probabilistic behavior of electrons within their orbitals, all contribute toward the stability of electron orbitals around the atomic nucleus. Further, these quantum characteristics (or Rules of Order) are not only *necessary* for the stability and the electromagnetic functionality of the Atomic System, but, as discussed later, are also *necessary* to enable covalent chemical bonds.

Time's Arrow and the Atom

Quantum processes and interactions are time-symmetric and reversible. Time at the quantum level can move backwards just as ready as forward. One consequence of this is that Atoms do not age. An Atom of Carbon is identical in every way to one that existed 1,000 million years ago, while objects made from such Atoms have dramatically and irreversibly changed (and decayed) over time. The same is true for all Atoms (except perhaps those involved in nuclear processes).

Time in the macro world only moves in the forward direction. This forward directionality of Time is commonly referred to as Time's Arrow. Physics has no valid explanation for this strange and fundamental behavior of Time. In particular, there is no good answer to Loschmidt's Paradox (first published in 1874), which states that it is not possible to deduce a time-irreversible process from a time-symmetric one.

What if Atoms did age? The universe could not function.

Further, Atoms do not mutate or evolve. Any changes to the Atomic Nucleus during a nuclear reaction are repeatable and predictable. The same 118 Elements that exist today existed thousands of millions of years ago. The Periodic Table never changes.

It might be instructive to briefly discuss the difference between random biological mutations and the changes in an Atom caused by nuclear reactions (including radioactivity). Evolution has a simple premise. Biological systems undergo random mutations over time. The mutations that enhance the survivability of that biological entity are incorporated over time, while the others fade away. For example, birds may develop longer beaks, or different color feathers, to adapt over time to a changing environment. In contrast, nuclear reactions are repeatable and predictable, and the Periodic Table never changes over time (no new Elements are created). Further, an Atom of say Carbon or Oxygen or Zinc is identical to one that existed billions of years ago. Atoms do not mutate.

Atoms are *stable and fundamental building blocks* that cannot be changed over Time (except during nuclear interactions). These characteristics are crucial for the functioning of the Universe.

Atomic Functionality and Purpose

The Atomic System is the fundamental building block of the universe whose structure and interactions produce multiple levels of functionality that enable the functioning of the universe, and everything within it.

The physical structure of the universe, including the Earth, would not be possible without the Atom. Atoms are the fundamental building blocks that enable the creation of physical entities such as stars and planets, as well as mountains, rocks, oceans, rivers, and clouds.

Subatomic particles within the Atom are the only particles to harness Electrical Charge, and result in the creation of all Electricity and Magnetism. These two properties also result in the creation of all Electromagnetism, and without which the universe could not function.

The Atom produces Electromagnetic Radiation (including Light, Radio waves, and X-rays) when an Electron drops to a lower-energy orbital within an Atom. The wave nature of electromagnetic radiation is evident from phenomena such as Diffraction and Interference, while its particle nature is evident in the Photoelectric Effect and in Blackbody Radiation. In the latter case, without such particle-like behavior, classical theories such as the Rayleigh-Jeans law, while correctly predicting radiated intensities at higher wavelengths, would incorrectly predict infinite total energies as the intensity diverges to infinity at the lower wavelengths.

The unique structure, composition, and behavior of the particles within the Atomic System enable the creation of the entire Periodic Table of the Elements, as well as the resulting properties of elements within the Periodic Table. A change in simply the number of protons (and neutrons) within the nucleus results in the creation of all the different elements within the universe, each with markedly different properties such as electrical and thermal conductivity. The Atomic System has the ability to combine different numbers of a few elementary subatomic particles to produce all the elements in the universe that are as diverse as Hydrogen, Argon, Carbon, Phosphorus, Silicon, Iron, Gold, and Sodium.

Further, electromagnetic emission spectra from the simpler atoms such as Hydrogen and Helium include only a few discrete bands of the much broader electromagnetic spectrum. The proton/neutron structure of the atomic nucleus enables heavier atoms such as Nitrogen and Iron to have larger numbers of electrons orbiting their nuclei. These electrons have many more high-to-low energy transition states that help produce the entire electromagnetic spectrum, without which the universe as we know it could not effectively function.

This unique structure of the Atomic System also enables the formation of both Ionic and Covalent chemical bonds that are responsible for producing chemical and molecular bonds, and all the resulting chemicals and compounds in the universe. The chemical properties of an element largely depend on the number of electrons in the outermost shell (valence electrons); atoms with different numbers of occupied electron shells but the same number of electrons in the outermost valence shell have similar properties, which gives rise to the periodic table of the elements.

In Ionic Bonds, the atoms lose or gain electrons, which then create a charge imbalance allowing oppositely charged ions to combine.

A Covalent Bond is a chemical bond between the Atoms within a molecule, and represents yet another aspect of quantum behavior necessary to enable one of the key functionalities of the Atom. Outer electrons and their orbitals are shared between Atoms, and exist in a superposition of their original orbitals as defined by Shrodinger's wave equation. The wave-particle duality of the electrons, as well as the probabilistic nature of electron distributions (as determined by the Shrodinger equation), are necessary for the creation of Covalent bonds.

Electrons orbiting the atomic nucleus cannot be described simply as solid particles. A more accurate analogy for electron orbitals might be that of a large and often oddly shaped "atmosphere" (the electron), distributed around the atomic nucleus. Atomic orbitals exactly describe the shape of this "atmosphere" only when a single electron is present in an atom. When more electrons are added to a single atom, the additional electrons tend to more evenly fill in a volume of space around the nucleus so that the resulting collection (sometimes termed the atom's "electron cloud") tends toward a zone of probability around the nucleus describing where the atom's electrons will be found. Without such behaviors, the Covalent Bond between Atoms and Molecules *would not be possible*.

The simplest case of a Covalent bond is the Hydrogen molecule with each of the two atoms having a proton as their nucleus. The probabilistic nature of electron distributions results in the creation of high energy density electrons between the two positively charged atomic nuclei to produce an attractive force that keeps them from repelling each other, resulting in a stable bond.

Molecular Orbital Theory is one the latest theories describing Covalent bonding. A discussion on this theory can be found in Ref. [6]. This reference includes a description of Slater-type orbitals, as well as Gaussian-type orbitals.

Covalent bonds between Atoms enable virtually all of Chemistry, including the formation of complex molecular chains such as carbohydrates, proteins, and the DNA molecule. This in turn enables all Biology, and all Life.

The atomic nucleus is another key aspect of the Atom and its functionality. It is crucial for multiple Protons to be able to bond together within the Atomic Nucleus. Without multiple Protons, more than one Electron could not simultaneously orbit the nucleus. Without the resulting structure of the Atomic System with multiple orbiting Electrons, it would not be possible for several elements to chemically bond with each other to create increasingly complex chemicals and compounds.

The forces within the atomic nucleus, when appropriately harnessed, allow for the creation of nuclear fusion and fission reactions. The unique structure and composition of the Proton and Neutron enables nuclear fusion, which enable the formation of stars, including our Sun, and without which life on Earth could not exist. These nuclear fusion reactions also enable the creation of heavier elements from lighter ones.

The Atom is the culmination of its quantum structure (quantized matter and force-carrying particles), its quantum properties (mass, quantized spin and quantized charge), quantum behaviors (quantized energy, wave-particle duality, quantum superposition, quantum uncertainty and quantum probability), and their interactions to enable the functionality of the entire Universe. The Atom incorporates virtually all the different facets of the quantum world. The Atom could be thought of as representing "Final Cause" that incorporates the structure, properties, and interactions of the quantum world to produce the structural, electromagnetic, nuclear, chemical and biological functioning of the universe.

The Atom is the *purposeful* coalescence of multiple *directed and independent* causes that work together to enable the entire functioning of the Universe. The Atom, in a loose sense, is God's Machine.

Underlying Mechanisms: The Elephant in the Room

The Standard Model of Physics has been developed over the past 80 years, and can accurately predict many aspects of quantum physics. The Standard Model is a non-abelian gauge theory with the symmetry group U(1)×SU(2)×SU(3) and has a total of twelve gauge bosons: the photon, three weak bosons and eight gluons.

However, Quantum Mechanics has no fundamental understanding of *why* matter exists as subatomic (quantized) particles in the first place. Or why they have a certain Mass, or why they have properties such as quantized Charge. Or *why and how* they exhibit both wavelike and particle-like behavior, and other quantum phenomenon such as Quantum Uncertainty and Quantum Superposition.

Instead, Quantum Mechanics begins with these fundamental experimental observations, and quantifies these observations using mathematical equations. For example, Edwin Schroedinger's Wavefunction, one of the foundational equations of Quantum Mechanics, is not based on any fundamental physics (unlike that describing an acoustic or a water wave that result from pressure and momentum considerations), but is semi-empirical in nature. Schroedinger simply came up with a wave equation that matched experimental observations of wavelike behavior (while conserving momentum and energy).

Further, much of Physics is based on the underlying symmetry of the Universe (although there are some key instances where this symmetry is broken). But there is no fundamental understanding of *why* the Universe is so beautifully symmetric. Physics once again assumes this is so based on experimental observations that have been repeatedly confirmed.

The Laws of Physics give us the *outcome* of subatomic particle interactions. However, they give us *no fundamental understanding* of the underlying mechanisms involved in these interactions.

For example, when we say that a particle such as an Electron has a negative Charge, what it really means is that *the particle has the ability to emit and to absorb massless Photons* that then travel at the speed of Light at a certain frequency and amplitude, and which exhibit wave-particle duality, and to thereby interact with other particles with Charge. Further, the mechanism by which a Photon, which in itself has no Charge, can transmit the electromagnetic force between charged particles remains unclear. This adds to the underlying complexity of the Photon.

The Electron, as it drops from a higher to a lower atomic orbital, emits a Photon. Further, an accelerating Electron emits Photons. The Electron has *the ability* to convert Mechanical Energy to Electromagnetic Energy. (This applies to all particles with electric Charge.)

Note that all matter consists of particles called Fermions, while forces between particles are mediated through particles called Bosons. Fermions include Electrons and Quarks, while Bosons include Photons and Gluons.

FERMIONS matter constituents spin = 1/2, 3/2, 5/2, ...

Leptons spin =1/2			Quarks spin =1/2		
Flavor	Mass GeV/c^2	Electric charge	Flavor	Approx. Mass GeV/c^2	Electric charge
ν_L lightest neutrino*	$(0-2)\times10^{-9}$	0	**u** up	0.002	2/3
e electron	0.000511	−1	**d** down	0.005	−1/3
ν_M middle neutrino*	$(0.009-2)\times10^{-9}$	0	**c** charm	1.3	2/3
μ muon	0.106	−1	**s** strange	0.1	−1/3
ν_H heaviest neutrino*	$(0.05-2)\times10^{-9}$	0	**t** top	173	2/3
τ tau	1.777	−1	**b** bottom	4.2	−1/3

Quantum attributes such as Mass, Spin, and Charge define and distinguish different subatomic particles. How does the particular combination of Mass, Spin, and Charge of the Electron *give it this ability* to emit and to absorb Photons, and to thereby interact with other charged particles? Physics has no fundamental understanding.

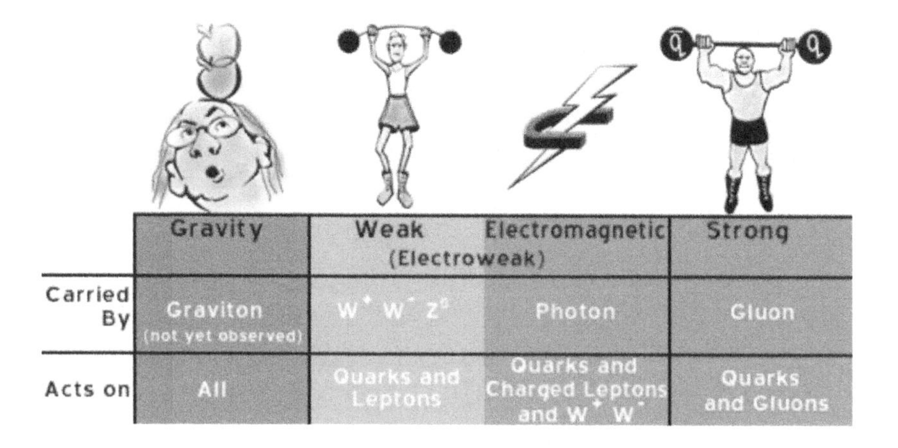

Similarly, a Quark has the ability to interact with other Quarks through the emission and absorption of different Gluons to produce the Color Force. Once again, how does the Mass, Spin, and Charge of the Quark *give it this ability* to interact with different types of Gluons? Again, Physics has no fundamental answer to these questions.

As mentioned before, quantum field theories such as QED and QCD can be used to describe such interactions, but fundamental questions about how such fields were created, and their ability to selectively interact with each other, still remain.

And could such interactions, and their underlying mechanisms, originate simply from random events?

Quantum Field Theory

Quantum Field Theory (QFT) is a gauge theory formulated to combine Quantum Mechanics with Special Relativity Theory, as well as to address the interactions of multiple subatomic particles. In this theory, time-varying and quantized energy fields in the form of quantized waves are postulated to pervade through space and time. Particles are represented as excitations of these underlying energy fields. Note that unlike QED and QCD, Quantum Field Theory treats not only the force-carrying particles (photon, gluons etc.) as excitations of an underlying field, it also treats the particles of matter (electron, quarks, etc.) in the same way.

QFT is the mathematical and conceptual framework for contemporary elementary particle physics. In a rather informal sense QFT is the extension of quantum mechanics (QM), dealing with Particles, over to Fields, i.e. systems with an infinite number of degrees of freedom.

However, the questions still remain: Why and how does Energy exhibit this time-dependent wavelike behavior? What gives the Fermions *the ability* to selectively interact with each other? What creates the quantization of these underlying fields and their interactions? Why do only certain of these fields interact with each other and not with others?

Quantum Field Theory results in a *mathematical representation* of subatomic particles (akin to how maps represent topology of the Earth's surface). Quantum Field Theory is an attempt to theoretically represent and match experimental observations of quantum behavior.

Reference [7] contains a comprehensive discussion regarding QFT from a philosophical viewpoint. From [7]: "A *particle interpretation* of QFT answers most intuitively what happens in particle scattering experiments and why we seem to detect particle trajectories. Moreover, it would explain most naturally why particle talk appears almost unavoidable. However, the particle interpretation in particular is troubled by numerous serious problems. There are no-go theorems to the effect that, in a relativistic setting, quantum "particle" states cannot be localized in any finite region of space-time no matter how large it is. Besides localizability, another hard core requirement for the particle concept that seems to be violated in QFT is countability. First, many take the Unruh effect to indicate that the particle number is observer or context dependent (the hypothetical Unruh effect is the prediction that an accelerating observer will observe blackbody radiation where an inertial observer would observe none). And second, interacting quantum field theories cannot be interpreted in terms of particles because their representations are unitarily inequivalent to Fock space* (Haag's theorem), which is the only known way to represent countable entities in systems with an infinite number of degrees of freedom.

At first sight the *field interpretation* seems to be much better off, considering that a field is not a localized entity and that it may vary continuously -- so no requirements for localizability and countability. Accordingly, the field interpretation is often taken to be implied by the failure of the particle interpretation. However, on closer scrutiny the field interpretation itself is not above reproach. To begin with, since "quantum fields" are operator valued it is not clear in which sense QFT should be describing physical fields, i.e. as ascribing physical properties to points in space. In order to get determinate physical properties, or even just probabilities, one needs a quantum state. However, since quantum states as such are not spatio-temporally defined, it is questionable whether field values calculated with their help can still be viewed as local properties. The second serious challenge is that the arguably strongest field interpretation—the wave functional version—may be hit by similar problems as the particle interpretation, since wave functional space is unitarily equivalent to Fock space." (*Note that the Fock space is an algebraic construction used in quantum mechanics to construct the quantum states space of a variable or unknown number of identical particles from a single particle Hilbert space H. The mathematical concept of a Hilbert space extends the methods of vector algebra and calculus from three-dimensional space to spaces with any finite or infinite number of dimensions).

Quantum Field Theories represent particles as oscillations that are the excitations of underlying fields. However, recent measurements on the dipole moments of subatomic particles by a team of scientists from Yale and Harvard to determine their sphericity [8] show no indication of any detectable internal oscillations in their dipole moments, and instead indicate that the particles are excruciatingly spherical in shape to less than a thousandth of a percent of their diameter.

Even if these internal oscillations existed, they would require a very complex spectral wave and amplitude distribution to realize a spherical shape. This could potentially violate the Uncertainty Principle, which under QFT postulates that the wavelength of the particle determines its momentum. A particle with a precisely-defined momentum would have a wave function with just a single wavelength, while one with a precisely-defined position would need a wave packet made up of many different wavelengths.

So, firstly, the measured sphericity of a particle would need a very complex distribution of waves. Secondly, even if verified, what underlying mechanism would *make* the Quantum Fields behave in such complex ways? And could such complexity arise simply from random events?

More recently (in 2018), Physicists from the Advanced Cold Molecule Electron Electric Dipole Moment (ACME) Collaboration have confirmed with very high precision that the charge of an Electron is excruciatingly spherical. This result, reported in the journal Nature [9], again confirms the Standard Model of Physics and discounts alternative theories such as QFT.

To quote Professor Gerald Gabrielse, a member of the ACME Collaboration: "If we had discovered that the shape wasn't round, that would be the biggest headline in physics for the past several decades. But our finding is still just as scientifically significant because it strengthens the Standard Model of particle physics and excludes alternative models." Note that the sphericity (and resultant symmetry) of subatomic particles is necessary for the time reversibility of subatomic interactions and of the Atom.

The hypothesis postulated by de Broglie very early on in which particles ride on an underlying wave field is being proposed by some physicists as an alternative. The debate continues on how to explain the wave-particle behavior, and the collapse of the wave function when an observation is made. Either way, Physics has *no fundamental understanding* of what produces these particle/wave behaviors.

Under a QFT formulation (setting aside its above-mentioned issues), Atoms are then a *purposeful* coalescence of numerous excitations of the underlying quantized (matter and force) fields that represent different subatomic particles, each having properties such as mass, quantum spin and quantum charge, and whose interactions provide the multiple levels of functionality necessary for the workings of the entire universe. The question still remains: *Could the existence of these numerous quantized force and matter fields and their selective interactions be simply the result of random events?*

Inexplicable Structure and Complexity

The following factors are necessary to enable the structure and the resultant multiple functionalities of the Atom:

1. Protons consist of 11 particles that dynamically interact with each other. It is surprising that these11 particles *selectively combined* with each other following the Big Bang, and especially since the two positively-charged up-Quarks in a Proton should have repelled each other?
2. Similarly Neutrons consist of 11 particles that dynamically interact with each other. It is again surprising that these 11 particles *selectively combined* with each other moments after the Big Bang, and especially as the two negatively-charged down-Quarks in a Neutron should have repelled each other?
3. The Proton-Neutron structure of the Atomic Nucleus is not possible without their complex structures and interactions. Two or more positively-charge Protons would repel each other, and increasingly so at the very small distances within the nucleus. The structure and interactions of the Protons and Neutrons with virtual Pions enables the creation of the residual strong nuclear force that overcomes the repulsive electrostatic force between the Protons. This structure of the Protons and Neutrons is necessary for the creation of the Atomic Nucleus.

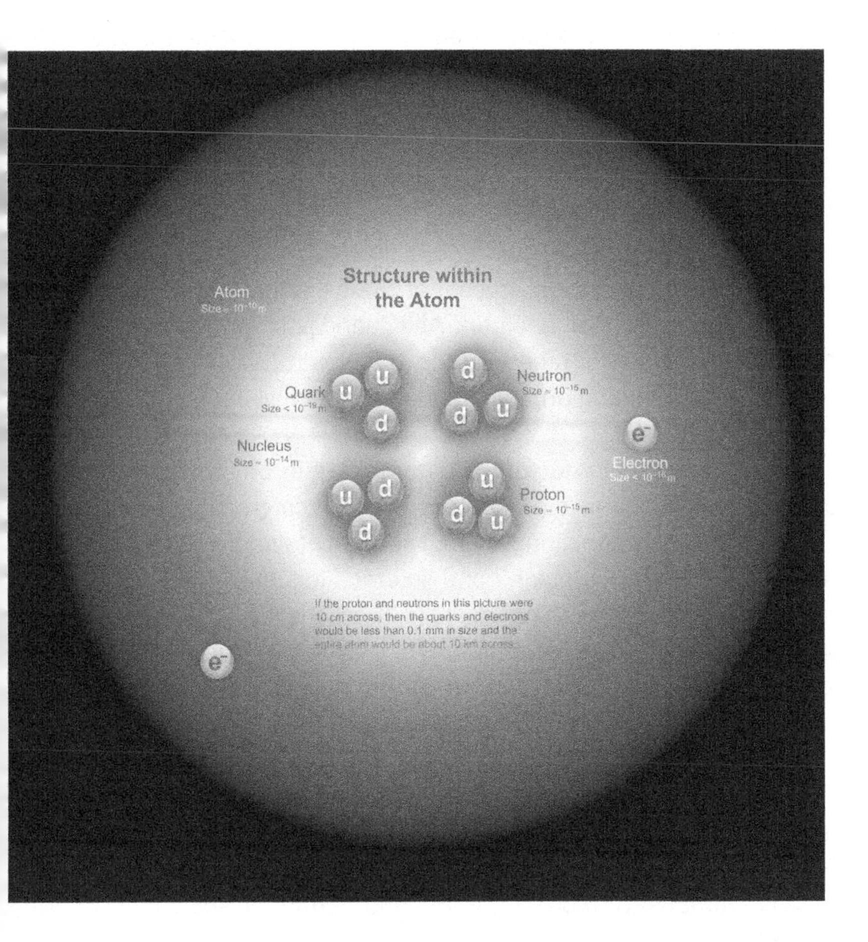

4. This structure of the Protons and Neutrons is also necessary in nuclear fusion reactions that fuel the stars, including our Sun.
5. The wavelike behavior of electrons, as well as those of quarks within the nucleons, is necessary for the stability and the electromagnetic functioning of the Atom.
6. Electrons produce quantized pulses of electromagnetic radiation when they change their orbit around an atomic nucleus.

7. The Electron has *the ability* to emit and absorb photons that enable the creation of the electromagnetic spectrum.
8. The Proton and the Neutron have *the ability* to emit and absorb gluons that enable the creation of the color and the residual nuclear forces.
9. The Proton and the Neutron have *the ability* to dynamically interact with each other through their interaction with virtual Pions to create the residual strong nuclear force that binds them together.
10. Pions consist of a Quark-AntiQuark pair, and appear out of Nothing to interact with the Protons and Neutrons within the nucleus. Normally, a quark and an antiquark should annihilate each other.

A multitude of *independent* factors are necessary to create the Atom with its Proton-Neutron-Electron structure, its dynamic interactions, its internal forces, and its resultant functionalities. The many factors listed above had to arise independently, combine, and interact together to produce The Atomic System with its resulting multiple functionalities. Further, unlike biological systems, Atoms do not mutate or evolve, and their structure and complexity cannot (in whole or in part) be explained by random mutations over time, so the Atom could be considered to be Irreducibly Complex.

The Why Questions related to the Atom and its causality are discussed in depth in the following section using Aristotle's theory of causality. Note that while the Atom is shown to be Irreducibly Complex, Aristotle's *Theory of Causality* is perhaps a more rigorous and formal assessment of the causality of the Atom.

The Why Question: Causality, Metaphysics, & Teleology

Discussion of causes and causality go back to Ancient Philosophy, featuring prominently in Aristotle's *Metaphysics* and *Physics*. Aristotle's theory of causality was developed to show that there are four (related) causes (or explanations) needed to explain change in the world. A complete explanation of any material change will use all four causes.

From [1]: "The study of nature was a search for answers to the question "why?" before and independently of Aristotle. A critical examination of the use of the language of causality by his predecessors, together with a careful study of natural phenomena, led Aristotle to elaborate a *theory* of causality. This theory is presented in its most general form in *Physics* II 3 and in *Metaphysics* V 5. In both texts, Aristotle argues that a final, formal, efficient and material cause can be given in answer to a why-question.

But Aristotle understood 'cause' in a much broader sense than we do today. In Aristotle's sense, a 'cause' or '*aiton*' is an explanatory condition of an object—an answer to a "why" question about the object. Aristotle classifies four such explanatory conditions—an object's form, matter, efficient cause, and teleology (the explanation of phenomena by the purpose they serve). In *Metaphysics*, an object's efficient cause is the cause that explains change or motion in an object. With the rise of modern physics in the seventeenth century, interest in efficient causal relations became acute, and it remains so today. And when contemporary philosophers discuss problems of causation, they typically mean this sense.

What is important is that this science consists in a causal investigation, that is, a search for the relevant causes. This helps us to understand why the most general presentation of Aristotle's theory of causality is repeated, in almost the same words, in *Physics* II 3 and in *Metaphysics* V 5. Although the *Physics* and the *Metaphysics* belong to two different theoretical enterprises, in both cases we are expected to embark on an investigation that will eventuate in causal knowledge, and this is not possible without a firm grasp of the interrelations between the four (types of) causes."

Aristotle's Metaphysics (which literally means "after the Physics") was the first major work in Philosophy to bear this name. This work deals with "the causes and principles of things", and is discussed in great detail in *Aristotle on Causality* [10], which points out that in this tradition of investigation, the search for causes was a search for answers to the question "why?"

From [10]: "In *Physics* and *Metaphysics*, Aristotle offers his general account of the four causes. Here Aristotle recognizes four types of causes that can be given in answer to a "why" question:

1. The Material Cause: "that out of which", e.g., the bronze of a statue.
2. The Formal Cause: "the form", "the account of what-it-is-to-be", e.g., the shape of a statue.
3. The Efficient Cause: "the primary source of the change or rest", e.g., the artisan, the art of bronze-casting the statue, the man who gives advice, the father of the child.
4. The Final Cause: "the end, that for the sake of which a thing is done", e.g., health is the end of walking, losing weight, purging, drugs, and surgical tools."

The Atom can be viewed as the confluence and the interaction of many fundamental aspects of quantum mechanics to realize the workings of the Universe. These include energy and its quantization into subatomic particles, with properties such as quantum spin, quantum charge & mass to produce matter and force-carrier particles, and quantum behaviors such as wave-particle duality, quantum superposition, the probabilistic nature of particle behaviors, and their time-symmetric interactions. These combine to enable the proton-neutron-electron structure to realize the structural, electromagnetic, chemical, nuclear, and biological functionalities of the Atom.

From Aristotle's viewpoint of causality, the four causes (or explanations) for the Atom in answer to a "Why?" question are:

1. Material Cause: Energy, with its ability to behave as both a wave and a particle. Energy was created at the Big Bang, the beginning of the Universe.
2. Formal Cause: The quantization of Energy into subatomic particles consisting of Matter- and Force-carrier particles (Fermions and Bosons) with quantized charge & spin, and with the Fermions having the ability to interact with each other by dynamically emitting and absorbing Bosons to create the stable Proton-Neutron-Electron atomic structure. (Note that the first atomic nuclei were created within moments after the Big Bang.) Other formal causes include the three color forces within the nucleus, the electromagnetic force between electrons and protons, and quantum behaviors of subatomic particles such as wave-particle duality, superposition, probability, and time reversibility. (Or, in the field formulation, the existence of the corresponding quantized force and matter fields and their selective interactions.)
3. Efficient Cause: An underlying Intelligence that produces and directs multiple independent causes (Material and Formal) to create the Atom with its complex structure and interactions. Most, if not all, of the properties, laws and behaviors of Quantum Mechanics, such as the quantization of Energy, the existence and quantization of charge and spin, the wave-particle duality of Energy, the existence & the dynamic interactions of the various atomic forces within and outside of the atomic nucleus, the time reversibility of the atomic interactions,

the probabilistic nature of particle distributions described by the Schroedinger wave equation, and the stability of the atomic system, are necessary to enable the multiple functionalities of the Atom.

4. Final Cause: The structural, electromagnetic, chemical, nuclear, and biological functionalities of the Atom to realize the workings of the Universe. The Atom makes the Universe function.

Aristotle's four causes together point to an underlying Intelligence as the only plausible explanation for the causality of the Atom. (Note also that the Atom is Irreducibly Complex). The Atom makes the Universe function. The Atom realizes the structural, electromagnetic, chemical, nuclear, and biological functionalities necessary for the workings of the Universe. Almost every aspect of the Atomic Structure and its quantized interactions is necessary to realize one of the many functionalities of the Atom. The Atom is the product of an underlying Intelligence that produces and directs multiple independent causes to enable its proton-neutron-electron structure with its quantum interactions to realize the workings of the Universe. The Atom must have a Designer (God).

Who Designed the Designer?

Does every Designer have a Designer? There are many things that we don't comprehend. For example, Physics does not have a *fundamental* understanding of how something could appear out of nothing, or how quantum objects can exist in multiple states *at the same time*. And as pointed out earlier, Quantum Physics does not have a fundamental understanding of many of the underlying mechanisms involved in its interactions. Similarly, it is difficult for us to fathom how something could always have existed outside of our time (and possibly space) and without having its own cause. Nonetheless, this does not mean it is not possible.

And while many things may be unfathomable in this world, and in Physics, we can at least show that our Universe *does* have a Designer.

The belief in a Creator God is well supported by the Big Bang Theory, which postulates that Space, Time and Matter all came into being temporally out of nothing right at the Big Bang. God, who always existed outside of our Space and Time, created the Universe. A well-reasoned argument for a Creator God can be found in Ref. [11].

From [11]: "...Many arguments claiming to prove the existence of God have been proposed throughout the centuries. A popular argument is that, since all effects come from causes, there must have been a "first cause" that is outside the material world—an "uncaused cause".

The response to many of these arguments, however, is: "If God created the world, what created God? In other words, if everything in the universe has a cause, why does God get a free pass? Don't we need an explanation for his origin as well?

In order to answer such questions, we first need to clarify what we mean by "God." If God is just another one of the causes within the system of causes that science explains, then we would need to search for a cause for God as well. But if God is something fundamentally different from the created order (what theologians call "transcendent"), then our demand for a cause of God's being is confused and misapplied.

God is not just the explanation for the beginning of the universe, but for the existence of anything at all—whether past, present, or future. These things are contingent; that is to say, they don't have to exist, and so because they do exist, we are right to ask for the causes of their existence. But theologians have understood God to be a necessary being. Asking for a cause of a necessary being is like asking how much the color blue weighs -- it is a category mistake.

The discovery in the past hundred years of strong evidence for a point of beginning for our universe (the "Big Bang") has had a tremendous impact on this discussion. Many have seen the "Big Bang" as proof that time, space, and matter are temporal, and not eternal -- which indeed point to the need for a creator. Note, however, that God would be the creator even if the universe did not have an empirically discernible beginning as some current theories (such as those concerning the "multiverse") suggest.

None of these theories can ever explain why anything exists in the first place. Science is powerless to answer that question, because it can only speak in terms of cause and effect. Every worldview must believe in a cause that itself is uncaused, and people understand this uncaused cause as the creator God,"

Note also that your kitchen table *does* have a Designer. This can be easily demonstrated by invoking Aristotle's four causes. One cannot then invoke an infinitely long series of Designers who in turn designed the succeeding one, to explain this Designer. Again, the only way the existence of a Designer would make sense is if there is an initial Designer that has always existed without a cause (outside of our Time, and possibly Space) from whom the other Designers originated.

The "Who designed the Designer?" question is also discussed in great depth in *The Last Superstition (2008)* by Edward Feser [11], and in *Who Designed the Designer? (2015)* by Michael Augros [13].

The Multiverse Hypothesis

The Multiverse is a controversial and hotly disputed idea that proposes the existence of an infinite number of Universes having their own natural constants and even their own physical laws. It is commonly cited as an argument against Design, as we then simply happen to be on the one Universe that has all the underlying factors necessary to support life.

A 2016 Smithsonian article by Sarah Scoles "*Can Physicists Ever Prove the Multiverse Is Real?*" points out that Astronomers are arguing about whether they can trust this untested - and potentially untestable - idea.

The Smithsonian article points out that the most common multiverse hypothesis is based on Inflation. "The universe began as a Big Bang and almost immediately began to expand faster than the speed of light in a growth spurt called "inflation." This sudden stretching smoothed out the cosmos, smearing matter and radiation equally across it like ketchup and mustard on a hamburger bun. That expansion stopped after just a fraction of a second. But according to an idea called the "inflationary multiverse," it continues - just not in our universe where we could see it. And as it does, it spawns other universes. And even when it stops in those spaces, it continues in still others. This "eternal inflation" would have created an infinite number of other universes."

Another version of the multiverse hypothesis is the "Many World's Interpretation". "Imagine, for example, that the cosmos is infinite. Then the part of it that we can see - the visible universe - is just one of an uncountable number of other, same-sized universes that add together to make a multiverse. Another version, called the "Many Worlds Interpretation," comes from quantum mechanics. Here, every time a physical particle, such as an electron, has multiple options, it takes all of them - each in a different, newly spawned universe."

But all of those other universes might be beyond our scientific ability to reach them. And because the multiverse is potentially unreachable, astronomers may not be able to find out if it exists at all. For an idea to move from hypothesis to theory, scientists have to test their predictions and then analyze the results to see whether their initial guess is supported or disproved by the data. If the hypothesis is consistently validated by experimental data, it gets recognized as an official theory.

The Smithsonian article continues: "Some theoretical physicists say their field needs more cold, hard evidence and worry about where the lack of proof leads. "It is easy to write theories," says Carlo Rovelli of the Center for Theoretical Physics in Luminy, France. Here, Rovelli is using the word colloquially, to talk about hypothetical explanations of how the universe, fundamentally, works. "It is hard to write theories that survive the proof of reality," he continues. "Few survive. By means of this filter, we have been able to develop modern science, a technological society, to cure illness, to feed billions. All this works thanks to a simple idea: Do not trust your fancies. Keep only the ideas that can be tested. If we stop doing so, we go back to the style of thinking of the Middle Ages."

He and cosmologists George Ellis of the University of Cape Town and Joseph Silk of Johns Hopkins University in Baltimore worry that because no one can currently prove ideas like the multiverse right or wrong, scientists can simply continue along their intellectual paths without knowing whether their walks are anything but random. "Theoretical physics risks becoming a no-man's-land between mathematics, physics and philosophy that does not truly meet the requirements of any," Ellis and Silk noted in a *Nature* editorial in December 2014.

...Sabine Hossenfelder of the Nordic Institute for Theoretical Physics in Stockholm, thinks "post-empirical" and "science" can never live together. "Physics is not about finding Real Truth. Physics is about describing the world," she wrote on her blog Backreaction. And if an idea (which she also colloquially calls a theory) has no empirical, physical backing, it doesn't belong. "Without making contact to observation, a theory isn't useful to describe the natural world, not part of the natural sciences, and not physics," she concluded."

Scientists may eventually be able to discover more direct evidence of the multiverse. They are seeking to find the stretch marks that inflation would have left on the cosmic microwave background radiation as evidence of collision between two or more universes.

"We are working on a problem that is very hard, and so we should think about this on a very long time scale," says UC Santa Barbara theoretical physicist Joseph Polchinski. That's not unusual in physics. A hundred years ago, Einstein's theory of general relativity, for example, predicted the existence of gravitational waves. But scientists could only verify them recently with a billion-dollar instrument called LIGO, the Laser Interferometer Gravitational-Wave Observatory."

Coles concludes: "So far, all of science has relied on testability. It has been what makes science and not daydreaming. Its strict rules of proof moved humans out of dank, dark castles and into space. But those tests take time, and most theoreticians want to wait it out. They are not ready to shelve an idea as fundamental as the multiverse - which could actually be the answer to life, the universe and everything - until and unless they can prove to themselves it *doesn't* exist. And that day may never come."

Many other physicists take a skeptical view of the multiverse hypothesis.

In a 2003 *New York Times* article, *A Brief History of the Multiverse,* Arizona State University professor and cosmologist Paul Davies offered a variety of arguments that multiverse theories are non-scientific: “For a start, how is the existence of the other universes to be tested? To be sure, all cosmologists accept that there are some regions of the universe that lie beyond the reach of our telescopes, but somewhere on the slippery slope between that and the idea that there are an infinite number of universes, credibility reaches a limit. As one slips down that slope, more and more must be accepted on faith, and less and less is open to scientific verification…”

George Ellis in 2011 [14] points out that the multiverse cannot be regarded as a traditional scientific theory. “As skeptical as I am, I think the contemplation of the multiverse is an excellent opportunity to reflect on the nature of science and on the ultimate nature of existence: why we are here.... In looking at this concept, we need an open mind, though not too open. It is a delicate path to tread. Parallel universes may or may not exist; the case is unproved. We are going to have to live with that uncertainty. Nothing is wrong with scientifically based philosophical speculation, which is what multiverse proposals are. But we should name it for what it is.”

Princeton cosmologist Paul Steinhardt stated his opposition to multiverse theories at the 2014 Annual Edge Foundation: “A pervasive idea in fundamental physics and cosmology that should be retired: the notion that we live in a multiverse in which the laws of physics and the properties of the cosmos vary randomly from one patch of space to another. According to this view, the laws and properties within our observable universe cannot be explained or predicted because they are set by chance. Different regions of space too distant to ever be observed have different laws and properties, according to this picture. Over the entire multiverse, there are infinitely many distinct patches. Among these patches, in the words of Alan Guth, "anything that can happen will happen—and it will happen infinitely many times". Hence, I refer to this concept as a Theory of Anything. Any observation or combination of observations is consistent with a Theory of Anything. No observation or combination of observations can disprove it. Proponents seem to revel in the fact that the Theory cannot be falsified. The rest of the scientific community should be up in arms since an unfalsifiable idea lies beyond the bounds of normal science. Yet, except for a few voices, there has been surprising complacency and, in some cases, grudging acceptance of a Theory of Anything as a logical possibility. The scientific journals are full of papers treating the Theory of Anything seriously."

Steinhardt theorizes that multiverse theories have gained traction mostly because too much has been invested in theories that have failed (e.g., Inflation Theory and String Theory). He sees in them an attempt to redefine the values of science, to which he objects even more strongly: "A Theory of Anything is useless because it does not rule out any possibility and worthless because it submits to no do-or-die tests."

Further, the Multiverse Hypothesis is based on the reasoning that quantum fluctuations following the Big Bang can create trillions of other universes. But this assumes that the differential equations that apply to our universe also apply to all the others, without first addressing how these differential equations originated in the first place. These differential equations resulted from theories and experiments on *our* Universe, but ***do nothing to explain how*** the structure, order and function (that these equations describe) of Atoms in our universe materialized to begin with from a quantum fluctuation.

The "anything is possible" Multiverse Hypothesis is essentially a cop out in trying to fathom and to explain the complexity and the mystery of our Universe.

Summary

All subatomic particles are ultimately made from Energy; but we have no fundamental understanding of what this Energy itself is.

We have no fundamental understanding of how Energy (Matter) is quantized into subatomic particles. Without this quantization of Matter, it would not be possible to create Atoms and the resulting functionality of the Universe.

Subatomic particles have inherent quantized properties of Charge and Spin, but we have no fundamental understanding of what Charge and Spin are. The conservation of Charge and Spin can be explained by the underlying symmetry of the physical laws governing their interactions; but this alone is not sufficient to fully explain what *enabled* the creation of Charge and Spin, or their characteristics. There appears to be no fundamental underlying explanation for Spin and Charge, properties that enable all electricity, magnetism, and electromagnetism.

Wavelike behavior results in stable electron orbits, and in the diffraction, reflection and interference of light. The symmetric nature of this wavelike behavior is necessary for the creation of the entire periodic table of the elements. Particle-like behavior manifests itself in the photoelectric effect, and can explain our observations of blackbody radiation where higher frequency radiation is limited to finite values, and which makes our Universe habitable.

The Atom can be viewed as the confluence of many fundamental (and independent) aspects of quantum mechanics to realize the workings of the Universe. These include energy and its quantization into subatomic particles, quantum properties such as spin, charge & mass to produce matter and force-carrier particles, quantum behaviors such as wave-particle duality, the probabilistic nature of particle behavior, time-symmetry, the Proton-Neutron-Electron structure of the atomic system with its dynamic interactions, and the resulting structural, electromagnetic, chemical, nuclear, and biological functionalities of the Atom.

The simplest atomic nucleus is a Proton consisting of two up-quarks and one down-quark that interact dynamically with eight gluons to create the three color forces. (Note that the exact number of gluons within a nucleon is murky and not definitively determined). This structure of the Proton is necessary for nuclear fusion reactions to occur, without which our Sun could not exist. It also enables the bonding of Protons with Neutrons, as well as the stability of neutrons within the nucleus. This structure also gives the proton a net integer positive charge that then enables Electrons to orbit the nucleus while precisely balancing out this charge.

Atoms are the fundamental platform for the production of the entire electromagnetic spectrum. The transition of an Electron from a higher to a lower orbital within an Atom produces a characteristic wavelength of radiation. Emission Spectra are produced by low-density gases that radiate energy at specific wavelengths characteristic of the element or elements that make up the gas. The Emission Spectrum consists of a number of bright lines against a dark background. Continuous Spectra are produced by solids, liquids or dense gases (all ultimately made up of a combination of numerous Atoms). The Continuous Spectrum appears as a smooth transition of all colors in the visible spectrum from the shortest or the longest wavelength without any gaps between the colors.

Neutrons, which are themselves made up of 11 subatomic particles, and which have no net electrical charge, have the ability to bond with one or more Protons. The interaction of Neutrons and Protons with virtual Pions produces the residual strong nuclear force that overcomes the inherent repulsive electrostatic force between positively-charged Protons to create heavier elements. This structure of the Neutron also supports nuclear fusion reactions that fuel our Sun, and which also results in the creation of heavier elements from lighter ones. The Proton/Neutron structure of the atomic nucleus also enables nuclear fission reactions that result from the splitting up of heavier nuclei into smaller ones. It also enables the creation of all the elements in the universe.

Further, it is crucial for multiple Protons to be able to bond together within the Atomic Nucleus for the creation of Chemistry (as well as for electromagnetism as discussed above). Without multiple Protons, more than one Electron could not simultaneously orbit the nucleus. Without the resulting structure of the Atomic System with multiple orbiting Electrons with their probabilistic wavelike behavior, it would not be possible for several elements to chemically bond with each other to create increasingly complex chemicals, compounds, and molecular chains (including carbohydrates, proteins, and the DNA molecule).

Atoms are a stable, time-reversible platform that does not age with time, and does not mutate or evolve. Any changes to the Atomic Nucleus during a nuclear reaction are repeatable and predictable. The same 118 Elements that exist today existed thousands of millions of years ago. The Periodic Table never changes. Without this stability and time reversibility of the Atom, the Universe could not function.

The following criteria are necessary for the creation of the Atomic System with its multiple functionalities:
Electrons have (1) a negative charge and (2) half-integer spin. They (and the (3) massless Photons they emit) also need the ability to exhibit (4) wave-like, (5) particle-like, (6) probabilistic and (7) time-reversible behaviors. Further, (8) they need to orbit and dynamically interact with the Atomic Nucleus (9) having a net positive charge and which has (10) the ability to combine different numbers of Protons and Neutrons. Protons

need to have (11) two up-Quarks and one down-Quark (12) interacting with eight Gluons (13) to produce three strong color forces and the resultant stored nuclear potential energy. Neutrons need to have (14) one up-Quark and two down-Quarks (15) interacting with eight Gluons (16) to produce a zero net charge. Neutrons are (17) stable only when combined with Protons. Further, (18) Protons and Neutrons dynamically interact with each other through (19) external virtual Pions consisting of a Quark – anti-Quark pair that (20) materializes from Nothing to (21) produce the residual nuclear force that enables multiple positively-charged Protons to combine together with Neutrons within the Atomic Nucleus.

Note that many of the above criteria occurred *independent* of each other. Note that unlike biological systems, Atoms do not mutate, so their complexity and functionality cannot be explained by evolutionary mechanisms. Atoms could be considered to be Irreducibly Complex.

Under a QFT formulation (setting aside its above-mentioned issues), Atoms are then a *purposeful* coalescence of numerous excitations of the underlying quantized (matter and force) fields that each represent different subatomic particles, each having properties such as mass, quantum spin and quantum charge, and whose interactions provide the multiple levels of functionality necessary for the workings of the entire universe. The question still remains: *Could the existence of these numerous quantized force and matter fields and their selective interactions be simply the result of random events?*

The Atomic System represents the "starting level" of complexity for the universe whose structure and interactions enable the physical, electromagnetic, nuclear, chemical, and biological functioning of the universe. Almost *every* aspect of this fundamental system, from its structure, to the properties, dynamic behaviors, internal forces, and dynamic interactions of its constituent particles, is necessary to enable one or more of its many functionalities.

Further, the heart of the Atomic System formed within moments after the Big Bang, and its complex structure and interactions cannot be explained using evolutionary mechanisms that require selective mutations over long periods of time to produce increasing complexity. And unlike biological systems, Atoms do not mutate or evolve. The Atom is the *purposeful* coalescence of multiple parts that dynamically interact to enable the entire functioning of the Universe.

The Standard Model has been repeatedly verified with great accuracy over the past 60 years, so it is fair to assume that any "new Physics" that addresses the effects of Gravity and of Dark Matter (which are currently not addressed by the Standard Model) should not significantly affect the Atomic System and its interactions; nor the conclusions reached herein. Note that Gravity is an extremely weak force in comparison to the other forces in the Standard Model, and Dark Matter can only be detected from its interactions with Gravity.

A common objection to the Designer argument is: "*Then who designed the Designer?*" There are many things that we don't comprehend. For example, Physics does not have a *fundamental* understanding of how something could appear out of nothing, or how quantum objects can exist in multiple states *at the same time*. And as pointed out earlier, Quantum Physics does not have a fundamental understanding of many of the underlying mechanisms involved in its interactions. Similarly, it is difficult for us to fathom how something could always have existed without a cause outside of our time (and possibly space). Nonetheless, this does not mean it is not possible.

And while many things may be unfathomable to us in this world, including in Physics, we can at least show that our Universe does have a Designer. This would perhaps be the case even in the unlikely event that the controversial Multiverse Hypothesis was someday experimentally validated.

So what do theologians mean by God? The belief in a Creator God is well supported by the Big Bang Theory, which postulates that Space, Time and

Matter all came into being temporally out of nothing right at the Big Bang. God, who always existed outside of our Space and Time, created the Universe.

God is not just the explanation for the beginning of the universe, but for the existence of anything at all -- whether past, present, or future. These things are contingent; that is to say, they don't have to exist, and so because they do exist, we are right to ask for the causes of their existence. But theologians have understood God to be a necessary being. Asking for a cause of a necessary being is like asking how much the color blue weighs -- it is a category mistake.

The discovery in the past hundred years of strong evidence for a point of beginning for our universe (the "Big Bang") has had a tremendous impact on this discussion. Many have seen the "Big Bang" as proof that time, space, and matter are temporal, and not eternal—which indeed point to the need for a creator. Note, however, that God would be the creator even if the universe did not have an empirically discernible beginning as some current theories (such as those concerning the "multiverse") suggest.

None of these theories can ever explain why anything exists in the first place. Science is powerless to answer that question, because it can only speak in terms of cause and effect. Every worldview must believe in a cause that itself is uncaused, and people understand this uncaused cause as the creator God,

Note also that your kitchen table *does* have a Designer. This can be easily demonstrated by using Aristotle's four causes. One cannot then invoke an infinitely long series of Designers who in turn designed the succeeding one, to explain this Designer. Again, the only way the existence of a Designer would make sense is if there is an initial Designer that has always existed without a cause (outside of our Time, and possibly Space) from whom the other Designers originated.

The Why Questions about the Atom are addressed using Aristotle's four causes. Aristotle's theory of causality was developed to show that there are four causes (or explanations) needed to explain change in the world. A complete explanation of any material change will use all four causes. From [8]: "In *Physics* and *Metaphysics*, Aristotle offers his general account of the four causes. Here Aristotle recognizes four types of causes that can be given in answer to a "why" question:

1. The Material Cause: "that out of which", e.g., the bronze of a statue.
2. The Formal Cause: "the form", "the account of what-it-is-to-be", e.g., the shape of a statue.
3. The Efficient Cause: "the primary source of the change or rest", e.g., the artisan, the art of bronze-casting the statue, the man who gives advice, the father of the child.
4. The Final Cause: "the end, that for the sake of which a thing is done", e.g., health is the end of walking, losing weight, purging, drugs, and surgical tools."

Note that Aristotle understood 'cause' in a much broader sense than we do today. In Aristotle's sense, a 'cause' or '*aiton*' is an explanatory condition of an object—an answer to a "why" question about the object. Aristotle classifies four such explanatory conditions—an object's form, matter, efficient cause, and teleology (the explanation of phenomena by the purpose they serve).

From Aristotle's viewpoint of causality, the four causes (or explanations) for the Atom in answer to a "Why?" question are:

1. Material Cause: Energy, with its ability to conform to both particle-like and wavelike behaviors. This Energy was created right at the Big Bang, the beginning of the Universe.
2. Formal Cause: The quantization of Energy into subatomic particles consisting of Matter- and Force-carrier particles (Fermions and Bosons) with quantized charge & spin, and with the Fermions having the ability to dynamically emit and absorb Bosons to create the stable Proton-Neutron-Electron atomic structure. (Note that the first atomic nuclei were created within moments after the Big Bang). Other

formal causes include the three color forces within the nucleus, the electromagnetic force between electrons and protons, and quantum behaviors of subatomic particles such as wave-particle duality, superposition, probability, and time reversibility. (Or, in the field formulation, the existence of the corresponding quantized force and matter fields and their selective interactions.)

3. Efficient Cause: An underlying transcendent Intelligence that produces and directs multiple independent causes (Material and Formal) to create the Atom with its complex structure and interactions. Most, if not all, of the properties, laws and behaviors of Quantum Mechanics, such as the quantization of Energy, the existence and quantization of charge and spin, the wave-particle duality of Energy, the existence & the dynamic interactions of the various atomic forces within and outside of the atomic nucleus, the time reversibility of the atomic interactions, the probabilistic nature of particle distributions described by the Schroedinger wave equation, and the stability of the atomic system, are necessary to enable the multiple functionalities of the Atom.
4. Final Cause: The structural, electromagnetic, chemical, nuclear, and biological functionalities of the Atom to realize the workings of the Universe. The Atom makes the Universe function.

Aristotle's four causes together point to an underlying transcendent Intelligence (God) as the only plausible explanation for the causality of the Atom. The Atom makes the Universe function. The Atom realizes the structural, electromagnetic, chemical, nuclear, and biological functionalities necessary for the workings of the Universe. Almost every aspect of the Atomic structure and its quantum interactions is necessary to realize one of the many functionalities of the Atom. The Atom is irreducible complex, and is the product of an underlying Intelligence that produces and directs multiple independent causes to enable its proton-neutron-electron structure with its quantum interactions to realize the workings of the Universe.

It is virtually impossible to explain the quantization, the structure, the stability and time reversibility, the internal quantum interactions, the internal forces, and the resultant multiple functionalities of the Atom from random events or through evolutionary mechanisms. The Atom is the product of an underlying transcendent Intelligence, and must have a Designer (God).

Note that while the Atom points to a Creator, it does not necessarily point to any particular religious belief. However, it does serve to reinforce the existence of God.

References

[1] *Metaphysics (2014),* Stanford Encyclopedia of Philosophy.

[2] *The Equivalence of Mass and Energy (2012)*, Stanford Encyclopedia of Philosophy.

[3] *Quantum Physics of Atoms, Molecules, Solids, Nuclei, and Particles* (2nd Ed.) by Robert Eisberg & Robert Resnick, 1985 (John Wiley & Sons, NY. pp. 252).

[4] *Symmetry and Symmetry Breaking (2003)*, Stanford Encyclopedia of Philosophy.

[5] Wigner, E. P., *Symmetries and Reflections* (1967), Bloomington, Indiana University Press.

[6] *Pictorial Molecular Orbital Theory*, Chemwiki, UC Davis (Chemwiki.UCDavis.edu)

[7] *Quantum Field Theory (2012)*, Stanford Encyclopedia of Philosophy.

[8] *Order of Magnitude Smaller Limit on the Electric Dipole Moment of the Electron (2013)*, Science Express.

[9] *Improved limit on the electric dipole moment of the Electron*, ACME Collaboration, *Nature* **volume 562**, pages 355–360 (2018). Published: 17 October, 2018.

[10] *Aristotle on Causality (2015)*, Stanford Encyclopedia of Philosophy.

[11] BioLogos: *What created God?* www.Biologos.org

[12] Feser, Edward, *The Last Superstition (2008),* St. Augustine's Press.

[13] Augros, Michael, *Who Designed the Designer?: A Rediscovered Path to God's existence (2015),* Ignatius Press.

[14] Ellis, George F. R. (August 1, 2011). *Does the Multiverse Really Exist?* Scientific American. New York: Nature Publishing Group. 305 (2): 38–43.

Throughout time, man has been in quest of security, of peace, and of happiness; however, rare are those who divine that their quest is in reality a quest for God.

- Fr. Paul Maria Sigl

Only a life lived for others is a life worthwhile.

- Dr. Albert Einstein

Acknowledgements

All images are from the Contemporary Physics Education Project (CPEP). Note that there is no implication of any sort that use of these figures implies any kind of endorsement or support of this book by CPEP.

Thanks also to my daughter Carey for creating and formatting the beautiful book cover.

Made in the USA
Monee, IL
16 September 2022